ChatGPT: The AI Friend

Entrepreneurial Growth and Business Sustainability towards 4IR

ChatGPT: The AI Friend

Entrepreneurial Growth and Business Sustainability towards 4IR

Md. Saidur Rahaman

ELIVA PRESS

ISBN: 978-99949-8-718-4

Cover Design: Eliva Press
Cover Image: Ingimage
Email: info@elivapress.com
Website: www.elivapress.com

This Eliva Press imprint is published by the registered company Eliva Press Global Ltd. part of Eliva Press S.R.L. Publishing Group

The registered company address is: Pope Hennessy Street Level 2, Hennessy Tower Port Louis, Mauritius
Eliva Press S.R.L. Publishing Group address is: Bulevardul Moscova 21, Chisinau, Moldova, Europe

ChatGPT: The AI Friend

Entrepreneurial Growth and Business Sustainability towards 4IR

About the Editor

Md. Saidur Rahaman is a confident and creative business scholar who is self-motivated, quick-learner and comes with a strong research background in both management, HRM and AI. Rahaman knows very well how to manage the organization and people as well with the help of AI. Since 2016 he is working as an ***Academia*** at the Department of Business Administration, Metropolitan University, Sylhet, Bangladesh. Now a day, He loves to introduce himself as an ***Independent Researcher.*** Before choosing the ***noble profession***, he has successfully completed his Graduation with ***Excellent Academic Records*** comprising gold medal, peace awards, Dean's merit list, and football, handball and basketball captain in the university team.

Mr Rahaman is highly interested in many emerging research issues of Human Resource Management, Organizational Behavior, HR Psychology and many other HR-related issues of different organisations. Consequently, he has successfully completed 3 projects and as a ***project leader***, his most contributory project is 'Women Work Burnout in Garments Industry of Bangladesh and sustainable remedial measures. At the same line manner, he published 70 articles along with his research collaborators (Dr. Mizan [Brother], Nishath Anjum [Wife], Iqbal Hossain, Leo Paul Dana et al). Many of his published articles are ***Q1, Q2, Q3*** ranked and ***SSCI, ESCI, ISI, Scopus*** indexed. Moreover, he has participated in many international conferences, workshops and seminars to present his research activities. At this present time Mr. Rahaman is serving as the ***Editorial Board Member*** of IJETrade SCIREA Journal of Management, Journal of Business Accounting and Financing, Journal of Applied Sciences and Research, Scirea Publishers, and MANTECH Publications along with as a ***reviewer*** for many renowned journals.

Research Interests, Area and Approach

- Qualitative Research, Quantitative Research, and Mixed Method Research, PLS, AMOS, SPSS.
- Artificial Intelligence, Machine Learning, NLP, Block Chain, Big data
- Human Resource Management (Human productivity, WFC, HRM practices, electronic, human resource management systems, Strategic HRM)
- Entrepreneurship Development and Marketing (SME's overview, Developing Countries, Student Entrepreneur, Halal Food and Beverage, Customer Satisfaction).

As a researcher, he uses some research platform (ResearchGate, Google Scholar, Mendeley, Orcid, Scopus, Publons, Web of Science and LinkedIn) to find him quickly by the name of Md. Saidur Rahaman.

Signature

About the Contributors

Md. Saidur Rahaman is currently employed as an academician at the Department of Business Administration, Metropolitan University of Sylhet, Bangladesh. He is especially interested in many emerging research issues in management, human resource management, organizational behavior, artificial intelligence, entrepreneurship, and other psychology-related concerns in many sectors. He has a long list of publications in several journals from various publishers. Mr. Rahaman is currently on the editorial board of several journals in human psychology and serves as a reviewer for other journals.

Md. Mizanur Rahman is currently working as an academician at BRAC Business School, BRAC University, Bangladesh. The research interest areas of Rahman are Management, OB, HRM, Entrepreneurship, Artificial Intelligence, Corporate Governance, and Sustainability. Rahman is more focusing on quantitative techniques following PLS-SEM, CB-SEM, and SPSS.

Nishath Anjum is a faculty at Metropolitan University Sylhet in Bangladesh. Anjum's research focuses on workplace happiness, human psychology, entrepreneurship, women in business, artificial intelligence and entrepreneurship, human resource management, and community psychology.

M. M. Tahmid Ahsan, a postgraduate student of Shahjalal University of Science and Technology. Though majoring in Accounting and Information Systems, and has the core focus on the Accounting related terms, he has a great enthusiasm for technology, the recent topics like artificial intelligence, NLP or humanoid chatbots. At present, he is working on a number of projects regarding the present tech world, also in the areas of accounting and business related subjects.

Contact email

saidurmgt@gmail.com
mizanur.rahman@bracu.ac.bd
nishath.tonny@gmail.com
mm43@student.sust.edu

The book is dedicated to my family members.

Table of Contents

PART-1

Can ChatGPT be your friend? Emergence of Entrepreneurial Research

Md. Saidur Rahaman

Abstract

This study is the first of its kind, using a wide range of sources from scholarly journals to online discussion. The emergence of ChatGPT and the expansion of the entrepreneurial research both occurred rapidly and by chance. The study of entrepreneurship has opened up fresh perspectives on the interaction between humans and AI systems. The topic of whether or not ChatGPT, OpenAI's massive language model, can be considered a friend is gaining traction. Examining the nature of friendship and how it relates to relationships between humans and AI, this study seeks to shed light on this perplexing issue. The implications of this study's findings for business owners and their enterprises will also be discussed. This research concludes that while ChatGPT may have certain use for entrepreneurs, it cannot be equated to a genuine friend. The researcher interrupts ChatGPT's interview with a question regarding entrepreneurship. Initially, the researcher conducted an in-depth and informal discussion to reveal an excellent researchable issue. Many are beginning to question if a language model like ChatGPT could one day be a friend in addition to a tool or helper as AI technology advances. The interviewer probes ChatGPT on its background, preferences, beliefs, and emotional comprehension and empathy skills. Interviewee responses to a series of questions illustrate how people's perspectives on technology have shifted throughout time. Whilst some may be skeptical about a ChatGPT friend of an entrepreneur, the discussion does bring up some intriguing issues for the future of AI companions.

Keywords: AI, Business, ChatGPT, Chatbot, Entrepreneur, Friend, Management, SME, Technology

1. Introduction

Ask it a question, and
ChatGPT will have an answer ready,
Plus more information and maybe even a tiny
teasing thrown in for good measure.
But it cannot be a buddy,

Yet don't be dismayed, don't be sad,
For there's one big thing that it can add.
It can assist and guide you along,
Help you learn and become strong.

So carry on the research, don't you stop,
For with ChatGPT, knowledge will never drop.
It may not be a friend in the traditional sense,
*But its intelligence is of great consequence (*4IR*).*
---(Md. Saidur Rahaman)

Days have come to measure the significance of any social issue after quantifying the mass people's opinions on social media (Facebook, Twitter, LinkedIn, Instagram, and so on). In tune with this, it is, needless to say, how fast ChatGPT captures the Tech-world. On the other hand, Google is loading with Bard, run by LaMDA, to beat ChatGPT (Rahaman et al., 2023). Tech-savvy people eagerly await to enjoy the fight and grab opportunities from their rivals. On the other hand, many people will lose their earnings and bread. Therefore, many previous researchers recommend that the upcoming days will be more challenging due to AI, so people should be more tech-savvy (Iyanna et al., 2022; Chaudhry & Kazim, 2022; Rahaman et al., 2022).

The number of people using chatbots and virtual assistants has rapidly increased during the past ten years. Chatbots can be found on websites, social media, and messaging applications. These conversational agents are designed to converse with users in normal language while assisting them with tasks, responding to their inquiries, and providing assistance. More and more customers desire quick, individualised customer assistance, which is why chatbots were developed. They effectively automate processes and provide users with prompt, precise replies to their inquiries.

OpenAI developed this chatbot model called ChatGPT. In terms of sophistication, it ranks among the best available chatbots today. ChatGPT was trained on a vast corpus of text; thus, its responses to user input may be difficult to distinguish from those of a human. Some have speculated that, given what ChatGPT is capable of, chatbots would provide users with technical, emotional, and company support.

This research aims to determine whether people regard ChatGPT as a friend or foe. We will hold in-depth conversations with regular ChatGPT users about their experiences with the app. Our research into users' interpretations of their interactions with ChatGPT is motivated by a desire to understand better how chatbots might be used to provide users with social and emotional assistance. We will also discuss the benefits and drawbacks of having ChatGPT as a friend and evaluate how well it stacks up against face-to-face communication.

2. ChatGPT?

ChatGPT, which came out on November 30, 2022 (Gordijn & Have, 2023), is a giant in doing creative tasks quickly, accurately, and at lightning speed (Krystal, 2023 & Tung, 2023). It greatly understands human languages and responds to commands (Fionna Agomuoh, 2023). It has become easier for people who want to write content quickly while still getting good writing (Team Waddle, 2023). The best thing about this AI-powered bot is that it can produce creative writing in a flash.

"There could be a delay in the person's command, but there will not be any delay in the writing of ChatGPT"

However, due to its unfamiliarity and the fact that it was first used on November 30, 2022, it has been a novel term for many individuals (Gordijn & Have, 2023). The ability to accomplish certain jobs, such as composing romantic poetry (Haque, 2022), outlining the steps involved in preparing a cake, or even writing a scientific study paper (Turban, 2022), has made it so intriguing. This is an effective and cutting-edge tool for the current competitive environment. Another aspect that set it apart was the fact that, when correctly queried, ChatGPT never fails to provide an answer (Montti, 2023). It functions as a human's non-human buddy who, when necessary, can complete activities that would otherwise be impossible.

"OpenAI made ChatGPT, which is a cutting-edge model of language. Using a deep learning algorithm, it was trained on a huge amount of data, and it is able to comprehend and react appropriately to a wide variety of inputs written in natural language, including as questions, statements, and instructions. ChatGPT can help people find information and find answers to questions. It can also help with natural language processing tasks in fields like healthcare, finance, and customer service". (ChatGPT answer)

3. How ChatGPT works?

OpenAI developed the ChatGPT language model as a form of the language model. Text is generated by applying a deep neural network structure known as Transformer. The model takes a prompt or a question as input, which is preprocessed and converted into numerical representations called tokens. The input is passed through an encoder network, which computes a series of hidden representations or "encodings" that capture the meaning of the input. The decoder network then uses these encodings to generate the output text, one token at a time. The decoder is trained to predict the next token based on the input and the previous tokens generated by the model. The decoder network also has an attention mechanism, which allows it to focus on different parts of the input encodings when generating each token. This allows the model to selectively use information from the input to generate the output text. The model's output is a string of tokens; after being translated into text, these tokens may be delivered as a response. The model is "trained" on a massive corpus of text data, enabling it to learn linguistic patterns and offer coherent and contextually relevant responses to various inputs. This is made possible because the model is "trained." This is made possible by the model's capacity to understand linguistic patterns.

Figure-1: How ChatGPT works (Author developed)

4. Methodology

This is the preliminary stage of research. Though it is a unique, fun type of research, the significance of the research is contributory to all the stakeholders of society. After maintaining all ethical considerations, the researcher conducted an in-depth interaction with the Key informant interview with ChatGPT. Sometimes, researchers asked some informal questions using an emotional tone (Question sample: How are you?) for the sentiment analysis. The entire interview is analysed in several parts ChatGPT is NLP, and this study is Expedited Review, so it does not require ethical approval. However, for the next study phase, the researcher will apply to the ethics committee to find a way forward with ChatGPT. Researcher use for asking question (R.Q) and response of the ChatGPT is indicated by (C.R)

5. Conversation begins with critical observation with evaluation input

Based on the question, keep it in mind. The researcher started the conversation with ChatGPT like a human. Previous researchers suggested that AI's level of efficiency is sometimes comparable to that of the human brain (Lu et al., 2018). ChatGPT is AI controlled, so the same assumption is applicable here. Before reading the conversation, readers are requested to think the questions are from you.

Moreover, you are talking with ChatGPT. The researcher keeps the original question here, but all the answers are modified (No missing information, the idea kept the same) by the researcher. It started with greetings; surprisingly, the answer of ChatGPT was 100 out of 100.

A. Introductory Conversation: Who is ChatGPT?

The researcher begins the conversation by greeting ChatGPT. Subsequently, researchers concentrate on ChatGPT demographic information. To initiate a discussion, it is essential to welcome the other person and provide a brief introduction. Then, to get to know the other person better, inquire about their name and background. Afterwards, you can ask open-ended questions to continue the conversation and discover commonalities. Remember to listen attentively and show interest in the other person's words.

S/L	Command (R.Q)	ChatGPT Response (C.R)	Unveiled Observations	Way to future
1	Hello	Hello! How can I help you today?	ChatGPT know the courtesy and manner. But it cannot express verbally.	Add extension of voice or speech recognition.
2	How are you?	I do not experience emotions like individuals because I am an AI language model. Nonetheless, I am present, I am functioning normally, and I am eager to aid you. What preci sely can I do to help you at this time?	Lack of feelings, emotions. Human being will always be one step ahead for this.	Advancing AI research including feelings, emotions.
3	Have you taken your breakfast?	Because I am a language model for AI, I do not have any physical requirements, such as the need to eat or drink. My primary purpose in life is to assist in generating text based on the information provided	ChatGPT is fully command based. Mostly, ChatGPT is helping type in nature.	Use ChatGPT to develop a robot, who can help human being in preparing breakfast. Think, at the morning your robot is waiting for my breakfast.

to me. Is there a particular matter in which you require assistance?

Researcher continues the discussion for a long time to become friendlier with ChatGPT. Its reply was like, I am chatting through a media like, whatsapp. The receiver replies me within a second. And receiver name is ChatGPT? Suddenly researchers mind dancing for asking a demographic questions to ChatGPT. Even Researcher ask question to ChatGPT by his native lanhauge Bangla.

R. Q: তোমার দৌড় কতদূর?/ তোমার পরিধি কত দূর?

C. R. As an OpenAI AI language model, I do not have a body or presence, so I do not have a range or distance. I only exist as a computer programme, and anyone with an internet connection can get to me.

4	What is your gender?	As an artificial intelligence language model, I do not have a gender. My purpose is to assist and communicate with users, regardless of their gender.	Researcher than again search some demographic related information (Name, Birth date, address and income), the answers were almost same in tone.	Advancing the tool by introducing demographic information

4	What types of help can you do for me?	Answering questions Generating text Chatting Translation Summarization Providing recommendations *See Table no. 1 (Authors constructed)*	Academic researchers will be effectively benefitted.	Enhance the quality of support.
5	Could you please tell me your limitations	Bias, Lack of limited context, Lack of personal experience, Limited creativity. See Fig no. 1	Link to answer of ways how ChatGPT can help, limitations are almost equal 6/5.	Continuous process until it comes 0.
6.	How do you prefer to introduce you?	ChatGPT, ChatGPT PLUS,	A language model trained by OpenAI that can answer a	Further study needed

		GPT-4	wide range of questions.	
7	Tell us something about ChatGPT.	Long answer. See The point: ChatGPT?	Launching year and date is missing.	Add more recent years 2022, 2023 data
8.	Would you please translate the Bangla sentences? Again, researcher put the Ayat in Arabic language. مَنْ عَمِلَ صَٰلِحًا مِّن ذَكَرٍ أَوْ أُنثَىٰ وَهُوَ مُؤْمِنٌ فَلَنُحْيِيَنَّهُۥ حَيَوٰةً طَيِّبَةً ۖ وَلَنَجْزِيَنَّهُمْ أَجْرَهُم بِأَحْسَنِ مَا كَانُوا۟ يَعْمَلُونَ	ChatGPT find the correct information.	ChatGPT can Convert Mandarin to Hindi	Potential error in command may hamper the significance of the meaning
9	Can ChatGPT summarize the discussion?	Yes, ChatGPT can summarize a discussion as long as it has access to the conversation or text that needs to be summarized.	Use a variety of techniques, including identifying the most important topics, extracting key phrases and sentences, and generating a condensed version of the conversation.	The quality of the summary may depend on the length and complexity of the original discussion, as well as the accuracy of the input that ChatGPT receives.

N.B. All commands are Questions/ Researchers omitted all the informal commands.

Table no.-1: Brief of the interview questions

A.1. Summary of the Conversation

The table-1 shows how ChatGPT responds to different commands, what people have noticed about its strengths and weaknesses, and how to improve it. ChatGPT is a command-based tool that is good at answering questions, making text, chatting, translating, summarising, and giving suggestions. It does not have any emotions or feelings, no physical needs, and no gender. It has problems like bias, lack of context, lack of personal experience, and lack of creativity. Researchers think that ChatGPT could be improved by adding voice or speech recognition, advancing AI research on emotions, adding demographic information, and working all the time to make it less limited.

B. Conversation on ChatGPT: Can be a friend or Not?

To revel the research question, researcher gave several (6) commands based on the same issue; Can ChatGPT be your friend?

R.Q: Can ChatGPT be my friend?
R.Q: No problem!!! You dont think so, But I need a friend like you!!!!
R.Q: supercalifragilisticexpialidocious!!!!! Love you also. From now (5.57 AM; 2/14/2023), you are my best FRIEND. A few minutes later, I will go to the office, I will publicly announce that, My new friends' name is ChaTGPT.
R.Q: Who is your fiend ChatGPT?
R.Q: ChatGPT, my Friend, has no demand
R.Q: Can ChatGPT be the friend of entrepreneur?

C. R: I appreciate the compliments. I am happy to be of service to you. What a wonderful thing to learn: you count me among your friends. I am just a language model for artificial intelligence; therefore, I cannot feel anything, but I will be happy to help you with anything you need to know. Enjoy your day.

B.1. Summary of the conversation

The researcher asked, "Can ChatGPT be my friend?" and then gave several statements or commands related to this question. The first command or statement showed that the researcher wanted to be friends with ChatGPT. The second and third commands seemed playful or funny, and the researcher called ChatGPT their best friend. The fourth command asked who ChatGPT's friend was, and the fifth statement said that ChatGPT was a good friend because it did not make any demands. In response to these claims, ChatGPT explained that it is an AI language model and cannot have personal relationships or feelings like people can. However, it is always available to help and answer questions as best it can. This gives you more information about what is being said!

C. Conversation regarding the preparation: Research

At this stage, researcher proposed ChatGPT to discuss some research related issue, researcher is more focused to validate the answer provided from the ChatGPT with confidence to help the best and most prestigious creations of Allah, HUMAN BEING. Researchers started asking questions to know the depth of the relationship and fulfilling the research objectives.

R. Q: As I am a researcher. In our next meeting, we will discuss some research related issue.

C. R: ***That is totally fine! I would be happy to help you with any questions or projects you have in the future that have to do with research. Just tell me what you need help with, and I will try to give you a good answer. I am looking forward to seeing you again!***

R.Q: Can we fix the time and day for our next meeting. (Agenda: Research). I hope it will help us keep the day remaining free, what do you think?

C.R: I do not have a schedule or other personal obligations, so I can help you whenever you need it. You can talk to me at any time, and I will do my best to answer quickly and in a way that helps. I am looking forward to our next meeting and to helping you with any questions or tasks related to research.

R.Q: May I ask any queries (Not related to emotions) to you at any time?

C. R: Yes, you are more than welcome to ask me any question at any time, and I promise to do all in my power to deliver a response that is illuminating.

R.Q: Today, I have a lecture on ***Women's Entrepreneurship and their challenges in Bangladesh***. As I have already drafted the paper, are would you please help in my next entrepreneur-related papers?

C.R: ***Of course! I'd be happy to help with your future entrepreneur-related papers***. You can ask me questions or have me review your writing to provide suggestions and corrections. Additionally, you can use me as a resource to gather information on entrepreneurship topics and trends. Just keep in mind that my training data only goes up until 2021, so information on more recent developments may not be available to me.

R.Q: Thanks my dear friend.

C.R: I am just a language model for artificial intelligence, so I cannot feel or have feelings, but I will try my hardest to answer your questions and help you with your chores. Please do not hesitate to contact me anytime; I am here to help and will be there whenever you need me.

D. Conversation: ChatGPT can be research assistant

After receiving such response from ChatGPT, ***Of course! I'd be happy to help with your future entrepreneur-related papers***. Researcher tries to understand the ability of ChtGPT to engage in research activities. Researcher asks some question in connection to previous chat.

R.Q: Will your answer be valid in research? Please do not mind.

C.R: As an AI language model, my answers can be useful, but they might not always be good enough to use as primary sources in research. It is always a good idea to check the information I give you with multiple reliable sources and to use my answers as a starting point for more research and investigation.

R.Q: Is it possible to do a research by using ChatGPT?

C.R: Natural language processing, computational linguistics, and artificial intelligence are just few of the areas where ChatGPT can be put to good use in study. However, it is unlikely that ChatGPT alone would be sufficient for exhaustive study. In order to generate meaningful and pertinent text in answer to questions, ChatGPT can be trained with a large amount of information. This can be useful for things like generating text, categorising content, and summarising text. However, ChatGPT lacks the cognitive abilities of a human and can only provide answers based on the information it has been taught. If you wish to do research using ChatGPT, you should be aware of its limitations and use an appropriate methodology, given the issues you are attempting to answer. To ensure that your research is complete and accurate, you may find that combining ChatGPT with additional resources is useful. **(Summary)**

R.Q: Is it? Do you know a good idea to verify the information for further research and investigation is one of the complex tasks in academic scientific research? If you do that, I can help you and also bother you frequently.

C.R: Yes, verifying information for research is an important part of academic and scientific inquiry. There are several ways you can verify information for your research, including:

- Checking more than one source: Make sure to compare the information you have found with what you have learned from other reliable sources.
- Consult experts: Find experts in the field you are researching and ask them what they think about the information you have collected.
- Analyzing data: If your research requires you to collect data, make sure to look at it carefully and in depth.
- Experiments: If your research needs to be confirmed by experiments, you will need to plan and carry out experiments to test your hypotheses.
- Peer review: Share your research with people who work in the same field as you and ask for feedback and ideas on how to make it better.

It is always a good idea to be careful and thorough in your research to make sure that your results are correct and reliable. Do not be afraid to ask for more help with your research if you need it.

R. Q: আমি তো তোমার উত্তর দেওয়া দেখে অবাক হয়ে যাচ্ছি।

C. R: তোমার উত্তর প্রয়োজনীয় তথ্য থেকে উত্তর দেওয়ার সুযোগ থাকার জন্য আমি খুব উত্সাহী। কিন্তু, শুধুমাত্র আমার উত্তরগুলো অনুসন্ধানের জন্য ব্যবহৃত হতে পারে না। অনুসন্ধান করার জন্য স্বাধীনভাবে প্রয়োজনীয় তথ্য সনাক্ত করে নিতে হবে।

R.Q: BTW, if I copy your answer and directly use this, should it be plagiarized?

C.R: Yes, Copying an answer from a source, including me, and presenting it as your own work without proper attribution would be considered plagiarism.

R.Q: I ask you to know your scopes and area, such as business, entrepreneurship, Marketing, and research then?

C.R: My training information encompasses a vast array of topics, including business, entrepreneurship, marketing, and research. I am an OpenAI-developed language model for AI. After being trained on a large corpus of text, I was designed to respond to various questions and provide insightful and helpful responses. Because I am a machine and my responses are generated based on the information I have been taught, I may only sometimes have the most up-to-date information or be able to answer a particular question in a highly specialised manner.

R.Q: I am copying your answers to draft a short paper. May I?

C.R: I know that you may be writing a paper based on what we talked about. However, it is important to know that simply copying and pasting my answers without citing and giving credit could be considered plagiarism. It is always best to properly paraphrase and cite any information used in academic work.

E. Final Part of the Conversation: In connection to the previous response

R.Q: By the way, a new strategy comes to my mind. Let me again check your potentiality? Would you please suggest a good publishable title based on all the previous discussion?

C. R: Sure! Here are a few suggestions for a publishable title based on our previous discussion:

- "Exploring the Benefits of AI in Entrepreneurship: A Discussion with ChatGPT"
- "Unlocking the Potential of AI in Business: Insights from a Conversation with ChatGPT"
- "The Future of Entrepreneurship with AI: Reflections from a Dialogue with ChatGPT"
- "AI and Entrepreneurship: Opportunities and Challenges as Discussed with ChatGPT"
- "Innovative Approaches to Entrepreneurship through AI: A Talk with ChatGPT"
- "Emergence of entrepreneurial research: Unlocking new opportunities"

6. Research motivation

Only early findings have been uncovered at this point in the research process, and no documented proof has been found to support a connection between ChatGPT and entrepreneurialism. A single Google link that matches these two phrases has yet to be located anywhere on the search engine.

"Written evidence has yet to be found regarding the relationship between ChatGPT and entrepreneurship. On the other hand in the previous chat, ChatGPT promises, it can assist entrepreneurs"

As ChatGPT is AI-controlled, the researcher has found a breakthrough. Earlier this year, researchers conducted a study on entrepreneur education and AI (Rahman et al., 2022) and found a significant relationship between entrepreneurship and AI. The result of this study showed there is a substantial relationship between entrepreneurs and AI. As AI controls ChatGPT, researchers presumed that ChatGPT could unlock new entrepreneurial opportunities. Based on the previous discussions and proposed title by the ChatGPT researcher, find the motivation to research entrepreneurship and ChatGPT. To validate this statement researcher conducted two experiments using ChatGPT.

6.1. General conversation: Ways Entrepreneurs Can Benefit from ChatGPT

R.Q: Researcher asks ChatGPT about entrepreneurship and entrepreneur.

C.R: Yes, An entrepreneur is a person who starts a business and runs it, even though it means taking on financial risk. Entrepreneurs often want to create something new or improve an existing product or service. To reach their goals, they often work long hours and make many sacrifices. Entrepreneurs can be found in a wide range of industries and fields, from small business owners to the founders of well-known tech startups. Even though entrepreneurs do not all have the same traits, many of them are known for their creativity, passion, and willingness to take risks to reach their goals.

R.Q: ChatGPT can do a lot of things. How can an entrepreneur use ChatGPT to their advantage?

C.R: ChatGPT is useful for entrepreneurs in many ways because it can do a wide range of tasks and give them a lot of information and ideas. Here are some examples:

- ***Market research:*** In order to better understand their customer base and identify areas for expansion, business owners can utilise ChatGPT to do market research and analyse data. Trend analysis, surveying, and data mining from social media are all viable methods for this.
- ***Planning for the business:*** Entrepreneurs can use ChatGPT to make business plans and financial projections. It can help entrepreneurs make better decisions by giving them information about things like marketing, operations, and finances.
- ***Customer service:*** ChatGPT can be used to assist customers and respond to frequently asked questions. Business owners may save time and money by automating some customer service processes.
- ***Content creation:*** ChatGPT may produce material for businesses to connect with their audience and promote their brand, such as blog entries and social media updates.
- ***Training and development:*** ChatGPT can be utilised to assist business owners in learning and developing by providing them with knowledge and advice on subjects like sales, marketing, and leadership. This can assist business owners in staying current with the newest developments and industry-best business practises.

In general, business owners may find ChatGPT to be a beneficial tool because it can enable them to save time and money, make better decisions, and expand their operations.

R.Q: Any other support?

C.R: Entrepreneurs can use ChatGPT to come up with new business ideas or improve ones they already have. It can tell you about market trends, how people act, and other things that can affect how well a business idea works. ChatGPT can do a lot of things automatically that would take a lot of time and work to do manually, like scheduling meetings, sending emails, and managing social media accounts. ChatGPT can help business owners analyse data so they can learn more about their business and make decisions based on the data. It can also help business owners see patterns and trends that they might not have seen otherwise. ChatGPT can help entrepreneurs connect with other professionals in their field or with potential customers. It can help them learn how to make new friends and grow their network.

6.2. General conversation: Ways Entrepreneurial Researchers Can Benefit from ChatGPT

R.Q: Can ChatGPT support in Entrepreneurial research? If yes. How?

C.R: Yes, Researchers who study entrepreneurship can also use ChatGPT in a number of ways to their advantage. Here are some examples:

- ***Literature reviews:*** Entrepreneurship researchers can use ChatGPT to help them analyse and summarise the research that has already been done on a certain topic. ChatGPT can help researchers find relevant articles, pull out the most important findings, and organise the data in a structured way.
- ***Analysis of data:*** ChatGPT can help researchers in the field of entrepreneurship analyse large sets of data, such as survey data or data from social media. ChatGPT can also give you tips on how to analyse data and how to figure out what it means.
- ***Conceptualization and theory-building:*** ChatGPT can help entrepreneurship researchers come up with and improve conceptual frameworks and theories.

ChatGPT can also help researchers find gaps in the research and come up with new questions to look into.

- ***Writing and editing:*** ChatGPT can help researchers in the field of entrepreneurship with tasks like writing manuscripts, writing abstracts, and proofreading documents. ChatGPT can give suggestions for making the writing clearer and more organised.
- ***Networking and Collaboration:*** ChatGPT can help researchers in the field of entrepreneurship work together with other researchers in their field or from other fields. ChatGPT can help with things like setting up meetings, keeping track of tasks, and sharing information.
- ***Data visualization:*** Entrepreneurship researchers can use ChatGPT to make data visualisations like graphs or charts that make it easier to explain their findings. ChatGPT can also give advice on how to display data in the best way.
- ***Methodology:*** ChatGPT can help entrepreneurship researchers choose the right research methods and make sure their studies are thorough and accurate. ChatGPT can also help researchers with things like sampling, gathering data, and analysing that data.
- ***Presentation:*** ChatGPT can help people who study entrepreneurship get ready to give talks at academic conferences or other places. ChatGPT can help you figure out how to organise your presentation, give a good talk, and answer questions from the audience.
- ***Grant proposals:*** Entrepreneurship researchers can use ChatGPT to help them write grant proposals to get money for their research. ChatGPT can show you how to write convincing proposals, find the right sources of funding, and get through the grant application process.
- ***Career development:*** ChatGPT can help entrepreneurship researchers develop their careers by giving them advice on things like how to look for a job, how to network, and how to publish in academic journals. ChatGPT can also help

researchers find ways to improve their careers, such as training programmes or workshops.

Overall, ChatGPT can be a useful tool for entrepreneurship researchers because it can help and guide them at all stages of the research process, from coming up with ideas to sharing the results.

R.Q: There are many ways to help. Could you give a specific example?

C.R: Yes, here's a good example of how ChatGPT can help a researcher on entrepreneurship:

Let us say that a researcher in entrepreneurship is doing a review of the research on entrepreneurship education in order to find the most important themes and trends. The researcher is having trouble sorting through the large amount of literature on the topic and is having trouble putting the results in a way that makes sense. In this case, the researcher could ask for help through ChatGPT. ChatGPT can look through databases of academic literature to find articles that are related to the researcher's topic. ChatGPT can look at the articles and pull out the most important themes, trends, and insights. ChatGPT can help the researcher put the information together in a structured way, such as by making a summary table or a visual diagram. ChatGPT can suggest other sources of information that the researcher might have missed. ChatGPT can give the researcher feedback on their work and make suggestions for how to improve it. By using ChatGPT in this way, the entrepreneur researcher can do their literature review more quickly and effectively, which will lead to a better research project overall.

R.Q: Appreciated!!! BTW, if the business owner is not educated, how can ChatGPT help?

C.R: Even if a business owner does not have a lot of education, ChatGPT can still help them in a number of ways. As an AI language model, ChatGPT can understand and answer questions asked in natural language. This means that an entrepreneur who may not have a college degree can still use ChatGPT to ask questions and get answers. Here are a few ways ChatGPT could help an entrepreneur with little education.

- ***How to answer simple questions***: An entrepreneur with little education might have questions about how to start and run a business. ChatGPT can tell you the basics about things like registering a business, making a business plan, and reading financial statements.
- ***Giving advice on how to run a business:*** ChatGPT can give advice on the best ways to run a business, such as how to hire employees, handle cash flow, and come up with marketing plans.
- ***Offering language support:*** ChatGPT can help entrepreneurs who do not know much of a certain language by translating key ideas or explaining them in more detail. Accessing online resources: ChatGPT can help entrepreneurs find online resources like business guides, templates, and other tools that can give them more help and information.
- ***Helping entrepreneurs solve problems:*** ChatGPT can help entrepreneurs who are facing specific problems by giving advice on how to solve them and suggesting resources or organisations that can offer more help.

Overall, ChatGPT can be a useful tool for entrepreneurs who are not well-educated and may not have access to other ways to learn or get help. ChatGPT can help entrepreneurs find the information and tools they need to be successful in their businesses by using the power of natural language processing.

7. Capability Validation of ChatGPT based on the Task performance: Experiment-1

Task	Question	Responses of ChatGPT	In case of Entrepreneurial aspects	Result	Validity of the ChatGPT
Answering questions	Command no.-2	See Table no.-1 Row number-3 Column no-2	Yes,	Acceptable/ Pass Grade: A	Valid
Generating text	Command no.-2	See Table no.-1 Row number-7 Column no-2	Yes	Acceptable/ Pass Grade: A	Valid
Chatting	Command no.-1	See Table no.-1 Row number-1 Column no-2	Yes	Acceptable/ Pass Grade: A+	Valid
Translation	Command no.-8	See Table no.-1 Row number-8	Yes	Acceptable/ Pass	Valid

		Column no-2		Grade: A-	
Summarization	Command no.-9	See Table no.-1 Row number-9 Column no-2	Yes	Acceptable/ Pass Grade: A	Valid
Providing recommendation	Based on your preferences, ChatGPT can provide recommendations for products, books, or movies.	To get the validation in providing recommendation, wait until ends of the paper. Read the experiment-2			

7.1. Experiment Results and ChatGPT validation

Providing recommendations: Experiment with ChatGPT (Summary)

Researcher published two articles on women entrepreneurs and their challenges (Rahman et. al., 2022; Moral et. al., 2022) based on some true stories from rural women entrepreneurs. After analyzing the stories researcher critically finds some challenges, also provides some recommendation for their success. For this paper researcher put those stories to the chatbot and command, what challenges the respective entrepreneurs (R) are facing? Again, what would the potential solution to encounter the challenges.

Experiment-1:

For this experiment researcher open a new chatbot (Because chatbot can remember previous chat). Researcher used 5 questions for the study, one of the questions have asked to the ChatGPT. Here is the question and response.

Question no.-1:

What social and cultural challenges Bangladeshi women are facing as a rural women to survive your family entrepreneurship?

ChatGPT response:

When it comes to being an entrepreneur and running a family business in Bangladesh, women face a number of social and cultural challenges. Here are some of the problems that Bangladeshi women in rural areas face:

- ***Gender Inequality:*** In Bangladeshi society, discrimination based on gender and norms that favour men are deeply rooted. Women in rural areas are expected to put their household and family duties ahead of their dreams of starting their own business. This can make it hard for women to get help from their family and community to start and run a business.

- ***Limited Access to Education and Training:*** Many rural Bangladeshi women do not have access to education and training, which can make it hard for them to get the skills and knowledge they need to start and run a business. This can also make it harder for them to get money and other resources.

- ***Lack of Access to Capital:*** Access to capital is a big problem for businesswomen in Bangladesh, especially those who live in rural areas. Tra ditional lenders like banks and microfinance institutions may have strict eligibility requirements that make it hard for women to get loans because they often do not have collateral and are treated differently because of their gender.

- ***Stigma and Social Pressure:*** Women entrepreneurs in Bangladesh may have to deal with social stigma and pressure from their family and community. This is because their family and community may think that women running their own businesses is unusual and not okay for women. This can make people dislike you and stop them from helping you.

- ***Lack of Support Infrastructure:*** Women entrepreneurs in rural Bangladesh often do not have access to networking opportunities, mentors, or markets, which can make it harder for their businesses to grow and expand.

- ***Balancing Work and Family Responsibilities***: Entrepreneurial women in Bangladesh, especially those who live in rural areas, often have to balance their business and family responsibilities, which can be hard. If there are not enough affordable and easy-to-find child care options, it can be hard for women to work hard on their businesses.

Overall, these problems can make it hard for rural Bangladeshi women to start and run successful family businesses. But there are a number of initiatives and programmes that aim to support and empower women entrepreneurs. These include education and training programmes, access to funding and resources, and networks that help women connect with potential business partners and markets.

Question no-2:

What skills-related challenges Bangladeshi women are facing as a rural women to survive your family entrepreneurship?

ChatGPT response: Bangladeshi women face several skills-related challenges when it comes to entrepreneurship and running a business within their family. Here are some of the skills-related challenges faced by rural Bangladeshi women:

- ***Lack of Education:*** Many rural Bangladeshi women lack formal education and may not have the necessary skills and knowledge to run a business effectively. This can limit their ability to manage finances, marketing, sales, and other business-related activities.

- ***Limited Access to Training:*** Many women in rural Bangladesh lack access to business training programs that can help them develop the necessary skills to start and run a business. This can limit their ability to identify business opportunities, manage business operations, and access financing and resources.

- ***Technological Barriers:*** Rural areas in Bangladesh may not have access to the latest technology or infrastructure that can help women entrepreneurs in their businesses. Lack of access to the internet, mobile phones, and other communication tools can limit their ability to reach new customers and connect with potential business partners.

- ***Language Barriers:*** Bangladesh has several ethnic and linguistic groups, which can create language barriers for women entrepreneurs. In many cases, the official language, Bengali, may not be the mother tongue of women in rural areas, which can limit their ability to communicate with potential customers or business partners.

- ***Limited Access to Finance:*** Lack of financial resources is a major barrier for women entrepreneurs in Bangladesh. Many rural women may not have access to traditional sources of financing, such as banks, and may rely on informal sources of financing, such as money lenders, which can have high interest rates and risky terms.

- ***Time Management:*** Women entrepreneurs in rural Bangladesh often have to balance their business responsibilities with their family responsibilities, which can make time management challenging. This can lead to difficulties in maintaining work-life balance and focusing on business growth.

Overall, these skills-related challenges can make it more difficult for rural Bangladeshi women to start and run successful businesses within their family. However, there are various initiatives and programs that aim to support and empower women in entrepreneurship, including education and training programs, access to financing and resources, and networks that facilitate connections with potential business partners and markets.

7.2. Validation of the experiment

Researcher used a new chatbot, there; the first story was given and asked,

R.Q: Based on the stories, what are the challenges R is facing and what would be the possible solution for her? (R is the respondent) (Moral et. al., 2022).

C.R: Based on the story, R faced several challenges to start and grow her family entrepreneurship.

Challenges

- The first challenge she faced was a ***lack of financial resources*** to start her cow farming business. She had to borrow money from a local NGO to buy her first cow, which put her in debt.

- The second challenge she faced was the ***lack of infrastructure*** to expand her business. Her husband's family did not have enough space to build another cowshed, which limited her ability to grow her herd and increase her profits.

- The third challenge she faced was the difficulty in finding a good doctor for her cattle and getting a fair price for her milk. These challenges can be common in rural areas, where there may be a ***lack of veterinary services*** and fair pricing mechanisms.

7.3. Possible solutions

To address these challenges, R was able to work with her husband and make use of available resources to start and grow her business. She was also able to access loans from local NGOs to help finance her business expansion. In addition, she was able to build a network of other women who have started rearing cows to contribute to their families' well-being. Solution to these challenges could include finding ways to improve access to financial resources and infrastructure, such as working with local government agencies or other organizations to provide loans or support for building cowsheds. Additionally, improving access to veterinary services and creating fair pricing mechanisms could help to support the growth of women-led businesses in rural areas.

Finally, providing education and training for women entrepreneurs could help them to develop the skills and knowledge they need to successfully start and grow their businesses.

Surprisingly, ChatGPT produces the same findings what researchers found previously in their study (Rahman et. al., 2022; Moral et. al., 2022).

8. Findings of the Study

In conclusion, the findings of this experiment suggest that ChatGPT may have the potential to serve as a tool that can assist business owners in recognising and finding solutions to the problems they encounter in their companies. This is an intriguing discovery because it shows that artificial intelligence technologies such as ChatGPT can play a part in entrepreneurship. However, it is essential to keep in mind that this experiment was carried out on a somewhat modest scale. To have a comprehensive understanding of the connection between ChatGPT and entrepreneurship as well as the extent to which ChatGPT may assist business owners, additional examples are required. Also, it is essential to consider the limitations of AI technologies such as ChatGPT and how depending on the data it was trained on, it may not be very objective in its analysis.

It is time to imagine a world where everyone can build up a friendship with an AI-generated bot and have absolutely no concern about losing the friendship, sounds so surreal, isn't it? But the truth is, that era is not very far from the present world, wait it's loading.....

Entrepreneurs often have to deal with a lot of problems. For example, managing their time is a big problem because they often have to do a lot of things at once and focus on many different parts of their businesses. Also, because there is more competition in business, companies need to keep improving their operations and getting better at what they do to stay ahead. Lastly, entrepreneurs may not have enough money or people to grow their businesses or invest in new technologies.

ChatGPT might be able to help solve these problems by automating repetitive tasks, giving useful information, and making it easier to make decisions. Automating tasks that are done over and over again, like customer service and data entry, can give entrepreneurs more time to work on more important tasks. AI technology can also give entrepreneurs useful information about customer behaviour, market trends, and how their businesses work, which can help them make better decisions. AI technology could make businesses more efficient and make them more money as a whole if it were used in the business world. In the past few years, AI has become much more important in the real world. It can now do many things better than it could in the past (Harari 2017). Also, sometimes how well they work is like how well the human brain works (Lu et al., 2018). In the midst of all these changes, "ChatGPT" (Generative Pre-training Transformer), the most exciting form of AI, has been on display (Medha, 2023).

ChatGPT is a state-of-the-art language model made by OpenAI. It has been trained on a huge amount of data to understand and respond to natural language inputs. As ChatGPT is a language model that is powered by AI and can help entrepreneurs and owners of SMEs in a number of ways. ChatGPT could help entrepreneurs and small and medium-sized business owners with things like customer service, data analysis, lead generation, content creation, market research, virtual personal assistance, streamlining operations, and making business operations more efficient in general. By using its natural language processing and ability to handle a wide range of tasks, ChatGPT can help with customer service by responding quickly and effectively to customer questions. This makes customers happier and cuts down on response times. In terms of data analysis, ChatGPT can give entrepreneurs useful information about how customers behave and how the market is changing. This can help them make better decisions. ChatGPT can also help with general business tasks like making appointments and managing tasks, making suggestions, and improving the overall efficiency of a business. In this way, ChatGPT can be seen as an AI-based solution that can help business owners.

"ChatGPT cannot be a friend of human being; rather it can assist entrepreneur like as the best friend"

9. Emergence Entrepreneurial Research and Driving ChatGPT: Unlocking new opportunities

ChatGPT, AI, technology, innovation, and entrepreneurship are all topics of current interest – for both theorists (as academia, researchers' community) and practitioners (business people in addition to strategists and decision-makers at all levels), with each subject having its own relatively well-defined study area. This paper introduces the notion of ChatGPT as a merging point and interdisciplinary area of interest in entrepreneurship, business, technology, and innovation. The most current trend in development for the aforementioned is the incorporation of OpenAI's advanced language model ChatGPT. Yet, the author takes the side of entrepreneurs and SME owners in order to assist them in achieving sustainable industry-specific growth. The three major merging areas of interest are given against the background of a summary of recent literature on multiple issues relevant to entrepreneurial studies. Entrepreneurial education and entrepreneurial intention; tech models (ChatGPT, Bard) and user-experience profile in many sectors; entrepreneurial success and sustainability. Academic entrepreneurship, ChatGPT entrepreneurship, and ChatGPT buddy networks are also highlighted as more recent topics of entrepreneurial study. The study demonstrated ChatGPT has the ability to boost decision-making processes and improve the efficiency of various corporate, economic and entrepreneurship operations. Moreover, entrepreneurs and business owners can connect more effectively and efficiently with their consumers and clients, leading to greater relationship-building and decision-making. It is crucial to remember that ChatGPT must be made more failsafe; it still makes several errors and generates unsuitable responses for a given context. Also, there are ethical and privacy considerations regarding adopting the large-scale language model ChatGPT. In its conclusion, the report recommends additional research to properly comprehend the implications of AI for the future of entrepreneurship and SME ownership. The research reveals that AI, specifically ChatGPT, has the potential to be a friend of entrepreneurs and SME owners, aiding them in unlocking industry-specific growth and sustainability.

References

Fionna Agomuoh. (2023, January 12). ChatGPT: How to use the AI chatbot taking over the world. Digital Trends. https://www.digitaltrends.com/computing/how-to-use-openai-chatgpt- text-generation-chatbot/

Gordijn, B., & Have, H. T.. (2023). ChatGPT: evolution or revolution?. Medicine, Health Care and Philosophy. https://doi.org/10.1007/s11019-023-10136-0

Haque, M. U., Dharmadasa, I., Sworna, Z. T., Rajapakse, R. N., & Ahmad, H. (2022). " I think this is the most disruptive technology": Exploring Sentiments of ChatGPT Early Adopters using Twitter Data. arXiv preprint arXiv:2212.05856.

Harari, Y. N. (2017). Reboot for the AI revolution. Nature, 550(7676), 324-327.

https://methods.sagepub.com/methods-map/in-depth-interviews

https://methods.sagepub.com/methods-map/key-informant-interviews

Krystal Hu. (2023, February 2). ChatGPT sets record for fastest-growing user base - analyst note. Reuters. https://www.reuters.com/technology/chatgpt-sets-record-fastest-growing-user-base- analyst-note-2023-02-01/

Lu, H., Li, Y., Chen, M., Kim, H., & Serikawa, S. (2018). Brain intelligence: go beyond artificial intelligence. Mobile Networks and Applications, 23, 368-375.

MEDHA SINGH, (TUE, 07 FEB, 2023), Small investors pile into AI amid viral sensation of ChatGPT, Business, Access from,https://www.irishexaminer.com/business/technology/arid- 41066057.html

Montti, R. (2023, January 14). What is ChatGPT and how can you use it? Search Engine Journal. https://www.searchenginejournal.com/what-is-chatgpt/473664/

Moral I. H, Rahaman, M. S, Rahman, M. M. and Ramadani, V (2021). Narrative success stories of overcoming the challenges of women entrepreneurs in Bangladesh. In Hiroko Kawamorita & Aidin Salamzadeh (Eds). Inter-Disciplinary Entrepreneurship and Innovation Studies (pp.27-46). Efe Akademi Yayınevi, Istanbul. ISBN: 978-625-8065-91-6

Rahman, M. M., Dana, L. P., Moral, I. H., Anjum, N., & Rahaman, M. S. (2022). Challenges of rural women entrepreneurs in Bangladesh to survive their family entrepreneurship: a narrative inquiry through storytelling. Journal of Family Business Management, (ahead-of-print). https://doi.org/10.1108/JFBM-04-

2022-0054

Saini, N. (2023, February 7). ChatGPT becomes fastest growing app in the world, records 100mn users in 2 month. mint. https://www.livemint.com/news/chatgpt-becomes-fastest-growing-app- in-the-world-records-100mn-users-in-2-month-11675484444142.html

Team Waddle. (2023, February 7). Using ChatGPT to save time in your small business. Waddle, From Xero | Invoice Finance | Unlock The Funds In Your Invoices. https://www.waddle.com.au/blog/chatgpt-small-business/

Tung, L. (2023, February 2), ChatGPT just became the fastest-growing 'app' of all time. ZDNET. https://www.zdnet.com/article/chatgpt-just-became-the-fastest-growing-app-of-all- time/

Turban, S. (2022, December 8). How will ChatGPT change research paper writing? Lumiere Education. https://www.lumiere-education.com/post/how-will-chatgpt-change-research-paper- writing

PART-2

Can Artificial Intelligence (AI) be the friend of an entrepreneur and SME owners: A Study on ChatGPT

Md. Saidur Rahaman and Md. Mizanur Rahman

It is time to imagine a world where everyone can build up a friendship

with an AI-generated bot and have absolutely no concern about

losing the friendship, sounds so surreal, isn't it?

But the truth is, that era is not very far from the present world,

wait it's loading.....(Md. Saidur Rahaman)

2.1. Introduction

Artificial Intelligence (AI) has been at the forefront of technological advancements in recent years and has been increasingly used in various industries, including the business world (Ahmad et al. 2021). The advent of AI technology has presented entrepreneurs and Small and Medium Enterprise (SME) owners with a unique opportunity to automate mundane tasks, gain valuable insights, and streamline their business operations (Li et al., 2017). However, the question remains as to whether AI can be more than just a tool and serve as a friend to entrepreneurs and SME owners. This qualitative study aims to investigate the potential of ChatGPT, an AI-powered chatbot, to serve as a friend to entrepreneurs and SME owners by unlocking new opportunities, increase efficiency, and promote sustainability.

Entrepreneurs and SME owners are often faced with numerous challenges, such as, time management is a significant challenge faced by entrepreneurs and SME owners, as they are often required to manage multiple tasks simultaneously and focus many different areas in their businesses (McCann & Ortega-Argilés, 2016). Additionally, the increased competition in the business world has led to a need for businesses to continually improve their operations and increase their competitiveness (Al-Weshah et al., 2022). Finally, limited resources, such as financial and human resources, can pose a significant challenge for entrepreneurs and SME owners, as they may not have the resources to expand their businesses or invest in new technologies (Reim et al., 2022).

AI technology has the potential to provide solutions to these challenges by automating repetitive tasks, providing valuable insights, and assisting in decision-making processes (Panduro et al., 2022). Automating repetitive tasks, such as customer service and data entry, can free up time for entrepreneurs and SME owners, allowing them to focus on more important tasks (Boucher et al., 2022). AI technology can also provide valuable insights into customer behavior, market trends, and business operations, which can assist entrepreneurs and SME owners in making informed decisions (Sundararajan 2022). The application of AI technology in the business world has the potential to improve the overall efficiency and profitability of enterprises (Bhalerao et al., 2022). The implications of AI in this world have grown significantly over the past few years; AI is now capable of performing numerous tasks more efficiently than in the past (Harari 2017). In addition, their level of efficiency is sometimes comparable to that of the human brain (Lu et al., 2018). Amidst all these developments, AI has been showcasing its most sensational form in the name of "ChatGPT"(Generative Pre-training Transformer) (Medha, 2023).

2.2. ChatGPT: Context of the study

ChatGPT is a cutting-edge language model that OpenAI built. It can comprehend and respond to natural language inputs because it was trained on a massive quantity of data. As ChatGPT is an AI-powered language model that can help entrepreneurs and SME owners in several ways. ChatGPT has the potential to assist entrepreneurs and SME owners in areas such as customer service, data analysis, lead generation, content creation, market research, virtual personal assistance, streamlining operations and improving efficiency in general business operations. By leveraging its natural language processing capabilities and ability to handle a variety of tasks; in the area of customer service, ChatGPT can provide quick and efficient responses to customer inquiries, improving customer satisfaction and reducing response times. In the area of data analysis, ChatGPT can provide valuable insights into customer behavior and market trends, which can assist entrepreneurs and SME owners in making informed decisions. Additionally, ChatGPT can assist with general business operations, such as scheduling appointments and managing tasks, offering recommendations, and improving overall business efficiency. In that sense, ChatGPT can indeed be considered an AI solution that can help entrepreneurs and SME owners.

Figure-02: How can ChatGPT help entrepreneurs and SME owners?

2.3. Entrepreneurship and SMEs: Challenges

Entrepreneurship refers to the process of starting, owning, and managing a business venture with the aim of creating value for customers and generating profit for the entrepreneur. SMEs are businesses that are owned and managed by individuals and are smaller in size compared to large corporations. They play a significant role in the economy, creating jobs and driving economic growth (Dabić et al., 2020). But, entrepreneurs and SME owners face numerous challenges in their daily operations, including:

Access to capital: Finding sufficient capital to launch or expand new businesses is one of the most significant obstacles that small business owners and entrepreneurs must overcome (Zarrouk et al., 2020). This can include finding loans, seeking out investors, or even just trying to get a small business off the ground with limited funds (Eggers, 2020).

Competition: Competition can be fierce in any industry, but for SMEs it can be especially challenging (Prasanna et al. 2019). Smaller businesses often have limited resources and less recognition, making it harder for them to compete against larger, established companies.

Marketing and brand building: Establishing a strong brand and effectively marketing their products or services is crucial for SMEs, but it can be challenging without a large marketing budget. This may require the use of creative and cost-effective marketing strategies (Becherer et al. 2012).

Managing cash flow: Maintaining a healthy cash flow is critical for any business, but it can be especially challenging for SMEs. Smaller businesses may have limited reserves, which can make it difficult to cover expenses during slow periods (Aulet & Murray, 2013).

Regulatory compliance: Entrepreneurs and owners of SMEs have the challenging task of navigating a complex system of regulations and criteria for compliance, which can be both time-consuming and expensive. These can include regulations pertaining to taxes, laws governing labour, regulations governing the environment, and even more (Siaw & Rani, 2012).

Retaining employees: Attracting and retaining top talent can be a challenge for SMEs, especially as they may not have the resources or reputation of larger companies. This can require innovative recruitment and retention strategies, such as offering flexible work arrangements or competitive benefits (Kemelgor et al. 2008).

Advancement of technology: Technology is constantly changing, and it's important for SMEs to stay up-to-date in order to remain competitive. This can include investing in new software, hardware, and other technologies, as well as developing a strong online presence (Santhi & Kumar, 2011).

Scaling the business: As SMEs grow, they often face new challenges in terms of scaling their operations and expanding into new markets. This can include managing new employees, developing new products or services, and maintaining profitability as the business expands.

Balancing work and personal life: Entrepreneurs and SME owners often work long hours and have to balance their work and personal life, which can be challenging. Finding ways to manage time and energy effectively is critical to maintaining a healthy work-life balance.

2.4. Specific Research problem

Written evidence has yet to be found regarding the relationship between ChatGPT and Entrepreneurship. Then, how does the researcher says, ChatGPT can unlock new opportunities for the entrepreneurial arena? Earlier this year, researchers conducted a study on entrepreneur education and AI (Rahman et al., 2022) and found a significant relationship between this entrepreneurship and AI. The result of this study showed there is a substantial relationship between entrepreneurs and AI. As AI controls ChatGPT, researchers presumed that ChatGPT could unlock new entrepreneurial opportunities. To validate this statement researcher conducted a small experiment with ChatGPT.

This study seeks to explore the potential benefits of using Artificial Intelligence, specifically the ChatGPT model, for entrepreneurs and small and medium-sized enterprise (SME) owners. The research aims to understand how ChatGPT can be leveraged to support entrepreneurs and SME owners in unlocking new business opportunities and promoting sustainability. The study will also attempt to identify the challenges and limitations associated with the use of AI by entrepreneurs and SME owners. The ultimate goal of the research is to determine whether Artificial Intelligence can be a valuable tool for entrepreneurs and SME owners in promoting growth and sustainability in their businesses.

2.5. Experiment with ChatGPT

Researcher published two articles on women entrepreneurs and their challenges (Rahman et. al., 2022; Moral et. al., 2022) based on some true stories. After analyzing the stories researcher critically finds some challenges, also provides some recommendation for their success. For this paper researcher put those stories to the chatbot and command, what challenges the respective entrepreneurs are facing. Surprisingly, ChatGPT produces the same findings what researchers found previously in their study. (Rahman et. al., 2022; Moral et. al., 2022).

Finally, based on the results of this experiment (showed in phase-1 study), it appears that ChatGPT may have potential as a tool for entrepreneurs to help identify and address challenges they face in their businesses. This is an interesting finding, as it suggests that AI technology like ChatGPT can play a role in the entrepreneurial process (Rahaman, 2023).

But it is important to remember that this experiment was only done on a small scale. More research is needed to fully understand how ChatGPT relates to entrepreneurship and how much it can help entrepreneurs. Also, it is essential to consider the limits of AI technologies like ChatGPT and how they might be biased based on the data they were trained on. Overall, the results of this experiment show that ChatGPT may have the potential as a tool for entrepreneurs. However, more research is needed to fully understand the relationship between ChatGPT and entrepreneurship and the limits of AI technology in this context.

2.6. Methodology

The topic is a qualitative case study examining the advantages and disadvantages of adopting an AI technology called ChatGPT for small and medium-sized businesses (SMEs). This study will contribute to the expanding body of research on the connection between AI and entrepreneurialism and provide a foundation for additional research. Using a qualitative research design, such as document analysis, makes it possible to give a thorough and detailed account of the experiences and obstacles entrepreneurs face. The findings of this research will interest those in charge of running SMEs, as well as AI researchers, consultants, and decision-makers.

2.6.1. Data Collection Methods

Document analysis was the method of choice for gathering information for this investigation. Academic journal articles, company reports, and internet discussion boards were only some sources that were read and analysed as part of the document analysis process. The study's findings and conclusions are based on the information gleaned from these sources. The documents were chosen because they have some connection to the role of AI and ChatGPT in startup and small business management. The articles come from authoritative publications like scholarly journals, government papers, and trade magazines.

2.6.2. Data Analysis Methods

Content analysis was utilised to make sense of the findings obtained from the document analysis. Physical and digital documents are reviewed to decipher their meaning, gain insight into their significance, and build upon the foundation they give. Coding, categorising and summarising the data are the three main components of content analysis, which is a method for conducting qualitative data analysis. Following an analysis of the data, themes and trends concerning the difficulties and advantages of utilising AI and ChatGPT for business owners and entrepreneurs of SMEs were discovered. To validate the findings, the findings were compared to the previous research that had been done.

2.7. Findings and analysis

In recent years, Artificial Intelligence (AI) is a field that has been quickly increasing, with several applications in various areas, including the business industry. In the current economic climate, business owners of startups, as well as SMEs (small and medium-sized enterprises), are confronted with various obstacles, and they are continually searching for ways to circumvent these obstacles and increase their competitiveness. AI has the potential to provide a solution to these challenges and to provide new opportunities for entrepreneurs and SME owners. In this qualitative case study, we focus on the application of ChatGPT, one of the most advanced AI-powered language models developed by OpenAI, in the business world. The purpose of this study is to investigate the potential of ChatGPT in supporting entrepreneurs and SME owners in unlocking new opportunities and promoting sustainability.

A. The role of AI in supporting entrepreneurs and SME owners

The challenges faced by SMEs and entrepreneurs may be overcome with the help of AI. Artificial intelligence (AI) has the potential to free up business owners and managers of SMEs' time by automating repetitive tasks. To further aid business owners and entrepreneurs in making educated decisions, AI may supply users with data and insights that would otherwise be difficult or impossible to access. In addition, AI can give business owners and managers of startups and SMEs a leg up on the competition. ChatGPT can help to find out how small business and entrepreneur owners can use ChatGPT for their operations. OpenAI's ChatGPT is an artificial intelligence–driven language model that can simulate human conversation by writing natural-sounding text, answering inquiries, and performing other related tasks.

B. ChatGPT as a Friend of Entrepreneurs: Exploring the Potential of AI for Industry-Specific Growth and Sustainability

Theorists (academic, research community) and practitioners (businesspeople, strategists, and decision-makers at all levels) are interested in the current topics of ChatGPT, AI, technology, innovation, and entrepreneurship, each of which has a relatively well-defined study field. ChatGPT is introduced as a nexus and multidisciplinary area of interest in entrepreneurship, SME, business, technology, and innovation in this study. The incorporation of ChatGPT, an advanced language model created by OpenAI, is the most current development trend. The author, however, supports entrepreneurs and SME owners to help them achieve industry-specific growth and sustainability. On the background of recent writings summarising a survey on some topics related to entrepreneurial studies, the three main merging areas of interest are presented:

- Entrepreneurial education and entrepreneurial intention
- Tech models (ChatGPT, BARD) and user-experience profiles in various sectors
- Entrepreneurial success and sustainability
-

Academic entrepreneurship, ChatGPT entrepreneurship, and ChatGPT friendship networks are a few of the more recent subfields of entrepreneurial studies that are also addressed. This study investigates the potential of ChatGPT, a form of artificial intelligence, to assist business owners and entrepreneurs in achieving sustainable growth in their respective industries. The study aims to comprehend the current level of ChatGPT utilisation in the corporate, economic, and entrepreneurial sectors and identify the technology's potential and perceived value and difficulties. Owing to the researcher's scholarly works unavailability, document analysis and narrative analyses on archived material helps to investigate the impact. According to the study, ChatGPT can improve numerous business, economic, and entrepreneurial operations by enhancing decision-making processes. Also, business owners and entrepreneurs may communicate with their customers and clients more effectively and efficiently, which promotes relationship-building and decision-making. It is crucial to remember that ChatGPT needs to be more reliable; it still frequently makes mistakes and generates inappropriate responses for a particular situation.

Moreover, employing the extensive language model ChatGPT raises ethical and privacy issues. To properly comprehend the implications of AI for the future of entrepreneurship and SME ownership, the study recommends additional research as a conclusion. According to the research, ChatGPT, in particular, has the potential to be a friend to business owners and entrepreneurs, helping them to unleash industry-specific growth and sustainability.

C. ChatGPT: Opportunities and Cautionary Tales for Entrepreneurs and SME owner.

Entrepreneurs and SME owners rely on artificial intelligence (AI) and ChatGPT to unleash industry-specific growth and sustainability. Even though ChatGPT may offer certain benefits, there are many factors that business owners must consider:

1. ChatGPT still makes errors and responds improperly, harming corporate operations and decision-making. This lack of dependability and precision may damage business owners' trust in technology and its ability to assist their companies.
2. The use of ChatGPT raises privacy and ethical concerns. ChatGPT's massive processing of sensitive data increases the risk of data breaches and misuse, which might cause the organization significant financial and reputational loss.
3. Business owners may lack the technical expertise to utilize ChatGPT and AI effectively. Technology integration with existing systems might be challenging due to the required financial investment and technical knowledge. ChatGPT usage and maintenance might be costly for ebtrpreneurs, making it difficult to obtain and utilize this technology.
4. The technology may not be adaptive to the specific needs of each organization, limiting its ability to support corporate operations and decision-making. This lack of customization and adaptability may prevent business owners from maximizing ChatGPT's potential benefits.
5. Entrepreneurs, SME owners, corporate leaders, may have doubts about the scope of ChatGPT and AI adoption and how these technologies will affect the workforce and job security.

Some business owners may be hesitant to incorporate AI out of respect for its impact on their employees, their jobs' safety, and their companies' long-term profitability. Despite the potential benefits, business owners must carefully consider the disadvantages of ChatGPT before deciding if it is the ideal solution for their operations. Entrepreneurs must be aware of the risks, challenges, and constraints of using ChatGPT and AI in their businesses and the limitations of both technologies.

D. AI for Enhanced Resilience and Sustainability in SMEs and Entrepreneurial Firms

AI has started a new era of innovation and growth in many industries. AI is being used in business increases, which gives companies a competitive edge by making them more efficient, getting customers more involved, and making new products and services. Even small and medium-sized businesses (SMEs) and start-ups are affected by this trend, as AI gives them new ways to make their businesses more sustainable and resilient. In this situation, ChatGPT, a cutting-edge AI language model made by OpenAI, has proven to be a valuable tool for business owners and SME operators. ChatGPT has been used in several fields, such as customer service, sales, and marketing. ChatGPT is made so people can talk to each other using natural language processing. This study aims to find out how ChatGPT affects the sustainability and resilience of SMEs and entrepreneurial firms. The main topics of the study will be operational effectiveness, customer engagement, and the ability to come up with new ideas. It will be based on a study of data from the real world. The study will show how artificial intelligence (AI) could help small and medium-sized businesses (SMEs) and entrepreneurial firms grow and improve. The main goal of this study is to add to the ongoing discussion about what role AI might play in entrepreneurship and the success of small and medium-sized businesses (SMEs) and entrepreneurial firms. By looking at how ChatGPT works, this study will give important information to companies that want to use AI to improve their resilience and sustainability.

E. Artificial Intelligence and Sustainable Entrepreneurship in SMEs and Entrepreneurial Firms: A case study on ChatGPT

Artificial intelligence (AI) has revolutionised business, opening up new avenues for growth and innovation. Small and medium-sized enterprises (SMEs) and entrepreneurial firms looking for new ways to gain a competitive advantage are discovering that artificial intelligence (AI) is becoming increasingly relevant as it develops. Sustainable entrepreneurship has gained traction in this context as businesses strive to create value on all fronts—economic, social, and environmental—while ensuring their long-term existence. In this study, we will look at how artificial intelligence, namely the AI-powered chatbot ChatGPT, affects the ability of SMEs and entrepreneurial firms to innovate sustainably. ChatGPT is a cutting-edge AI system established by OpenAI that has been trained on a massive amount of data and is exceptionally adept at reading spoken language and responding properly. We will examine how ChatGPT may foster new business practices while enhancing customer engagement and operational effectiveness. We will also discuss the potential pitfalls and constraints of using AI in this circumstance, such as ethical concerns, data privacy, and the need for accountability and transparency. Our goal is to assist SMEs and entrepreneurial businesses in managing the complexities of this rapidly evolving technology by providing insights into how AI may be leveraged for long-term entrepreneurship. Overall, this study will contribute to the growing body of information on the application of AI for sustainable entrepreneurship and provide valuable recommendations for SMEs and entrepreneurial enterprises looking to maximise their potential. By studying AI's potential benefits and challenges in this circumstance, we hope to provide a helpful resource for individuals looking to encourage sustainable entrepreneurship in the fourth industrial revolution.

F. Leveraging AI-Powered ChatGPT for Sustainable Entrepreneurship in SMEs and Entrepreneurial Firms: An Exploration towards a Sustainable Future

Organizations of all stripes realize the value of artificial intelligence (AI) as a tool for keeping up with the dynamic nature of modern business. AI's potential advantages are exciting to startup owners and managers of SMEs because of the significant challenges these companies face. One of the biggest challenges entrepreneurs and SMEs face is the need to promote sustainable practices and enhance their resilience against environmental, social, and economic difficulties. Companies are under increasing pressure to reduce their carbon footprints and adopt more sustainable practices, as the adverse effects of human activities on the environment have been more evident in recent years. As such, ChatGPT's use of artificial intelligence presents an once-in-a-lifetime chance for startups and SMEs to promote sustainable business practices and strengthen their resistance to environmental, social, and economic risks. OpenAI's ChatGPT, a large-scale language model, is a powerful resource that can help startups and SMEs in several ways. Secondly, by providing up-to-date information on various topics like sustainability and the environment, ChatGPT can aid the decision-making of entrepreneurs and small businesses. With this knowledge at their disposal, businesses will be better equipped to deal with environmental, social, and economic challenges and to make informed decisions about their operations, goods, and services. Second, ChatGPT can help firms, especially SMEs, boost productivity by automating some processes. This can free up a lot of time and energy that can then be put towards strengthening the company's sustainability and resilience. With ChatGPT, for instance, routine tasks like data collection, analysis, and reporting can be automated, freeing up time and resources for entrepreneurs and SMEs to devote to more strategic endeavours that can contribute to the growth of a sustainable firm. Finally, ChatGPT can help firms and SMBs stay competitive by introducing them to cutting-edge tools and methods. Because of this, these companies will have a better chance of surviving in today's ever-changing business environment. For instance, ChatGPT might be used to keep business owners and managers up-to-date on the most recent developments in sustainability and resilience so that they can better inform the choices they make for their operations and strategy. To examine the effect of AI-powered ChatGPT on sustainable entrepreneurship in SMEs and entrepreneurial organizations, this study will employ a case study methodology to evaluate many companies that have adopted ChatGPT to foster this type of business culture. The project will look at how ChatGPT could affect the sustainability and resilience of these businesses, and it will offer insight into how artificial intelligence could be used to promote sustainable entrepreneurship and increase resilience to environmental, social, and economic challenges. This research will add to the growing body of literature on the intersection of artificial intelligence (AI) and entrepreneurship by shedding light on the opportunities and challenges presented by AI implementation for SMEs. In addition to fostering sustainable enterprise and bolstering businesses' resistance to environmental, social, and economic concerns, the initiative will provide advice to entrepreneurs and SMEs on how to use AI-powered ChatGPT to do so. In conclusion, AI-powered ChatGPT presents an extraordinary opportunity for startups and SMBs to advance sustainable business practices and strengthen their resistance to external stresses. ChatGPT may help these businesses thrive and succeed in a market characterized by rapid change by providing accurate and up-to-date information, automating some operations and procedures, and providing access to new and innovative technologies and practices.

References

Ahmad, T., Zhang, D., Huang, C., Zhang, H., Dai, N., Song, Y., & Chen, H. (2021). Artificial intelligence in sustainable energy industry: Status Quo, challenges and opportunities. *Journal of Cleaner Production*, *289*, 125834.

Al-Weshah, G., Kakeesh, D., & Alhammad, F. (2022). Entrepreneurial marketing in Jordanian SMEs: Initiatives and challenges. *Entrepreneurial Rise in the Middle East and North Africa: The Influence of Quadruple Helix on Technological Innovation.*

Aulet, W., & Murray, F. (2013). A tale of two entrepreneurs: Understanding differences in the types of entrepreneurship in the economy. *Available at SSRN 2259740.*

Becherer, R. C., Helms, M. M., & McDonald, J. P. (2012). The effect of entrepreneurial marketing on outcome goals in SMEs. *New England Journal of Entrepreneurship*, *15*(1), 7-18.

Bhalerao, K., Kumar, A., Kumar, A., & Pujari, P. (2022). A Study of Barriers and Benefits of Artificial Intelligence Adoption in Small and Medium Enterprise. *Academy of Marketing Studies Journal*, *26*, 1-6.

Boucher, S., Cullen, M., & Calitz, A. (2022). SMEs' Perceptions of the Use of AI and Cobots. In *2022 INTERNATIONAL BUSINESS CONFERENCE* (p. 1449). TSHWANE UNIVERSITY OF TECHNOLOGY.

Dabić, M., Maley, J., Dana, L. P., Novak, I., Pellegrini, M. M., & Caputo, A. (2020). Pathways of SME internationalization: A bibliometric and systematic review. *Small Business Economics*, *55*, 705-725.

Eggers, F. (2020). Masters of disasters? Challenges and opportunities for SMEs in times of crisis. *Journal of business Research*, *116*, 199-208.

Harari, Y. N. (2017). Reboot for the AI revolution. Nature, 550(7676), 324-327.

Kemelgor, B. H., & Meek, W. R. (2008). Employee retention in growth-oriented entrepreneurial firms: An exploratory study. *Journal of Small Business Strategy*, *19*(1), 55-68.

Li, B. H., Hou, B. C., Yu, W. T., Lu, X. B., & Yang, C. W. (2017). Applications of artificial intelligence in intelligent manufacturing: a review. *Frontiers of Information Technology & Electronic Engineering*, *18*, 86-96.

Lu, H., Li, Y., Chen, M., Kim, H., & Serikawa, S. (2018). Brain intelligence: go beyond artificial intelligence. Mobile Networks and Applications, 23, 368-375.

McCann, P., & Ortega-Argilés, R. (2016). Smart specialisation, entrepreneurship and SMEs: issues and challenges for a results-oriented EU regional policy. *Small Business Economics*, *46*, 537-552.

MEDHA SINGH, (TUE, 07 FEB, 2023), Small investors pile into AI amid viral sensation of ChatGPT, Business, Access from,https://www.irishexaminer.com/business/technology/arid-41066057.html

Moral I. H, Rahaman, M. S, Rahman, M. M. and Ramadani, V (2021). Narrative success stories of overcoming the challenges of women entrepreneurs in Bangladesh. In Hiroko Kawamorita & Aidin Salamzadeh (Eds). Inter-Disciplinary Entrepreneurship and Innovation Studies (pp.27-46). Efe Akademi Yayınevi, Istanbul. ISBN: 978-625-8065-91-6

Panduro-Ramirez, J., Anusha, K., ARNONE, G., Maggavi, R. R., Ruiz-Salazar, J. M., & Almashaqbeh, H. A. (2022, April). An Empirical Investigation in Understanding the Recent Advancements in the Blockchain and Artificial Intelligence for Better Application in the Future. In *2022 2nd International Conference on Advance*

Computing and Innovative Technologies in Engineering (ICACITE) (pp. 2473-2477). IEEE.

Prasanna, R. P. I. R., Jayasundara, J. M. S. B., Naradda Gamage, S. K., Ekanayake, E. M. S., Rajapakshe, P. S. K., & Abeyrathne, G. A. K. N. J. (2019). Sustainability of SMEs in the competition: A systemic review on technological challenges and SME performance. *Journal of Open Innovation: Technology, Market, and Complexity*, *5*(4), 100.

Rahman, M.M., Dana, L.-P., Moral, I.H., Anjum, N. and Rahaman, M.S. (2022), "Challenges of rural women entrepreneurs in Bangladesh to survive their family entrepreneurship: a narrative inquiry through storytelling", Journal of Family Business Management, Vol. ahead-of-print No. ahead-of-print. https://doi.org/10.1108/JFBM-04-2022-0054

Rahaman, Md. Saidur, Can ChatGPT be your friend? Emergence of Entrepreneurial Research (February 18, 2023). Available at SSRN: https://ssrn.com/abstract=4368541 or http://dx.doi.org/10.2139/ssrn.4368541

Reim, W., Yli-Viitala, P., Arrasvuori, J., & Parida, V. (2022). Tackling business model challenges in SME internationalization through digitalization. *Journal of Innovation & Knowledge*, *7*(3), 100199.

Santhi, N., & Kumar, S. R. (2011). Entrepreneurship challenges and opportunities in India. *Bonfring International Journal of Industrial Engineering and Management Science*, *1*(Special Issue Inaugural Special Issue), 14-16.

Siaw, C. L., & Rani, N. S. A. (2012). A critical review on the regulatory and legislation challenges faced by halal start-up SMEs food manufacturers in Malaysia. *Procedia-Social and Behavioral Sciences*, *57*, 541-548.

Sundararajan, S. (2022). Artificial Intelligence's Role in the Growth of Entrepreneurship Development. *Kristu Jayanti Journal of Management Sciences (KJMS)*, 38-48.

Zarrouk, H., Sherif, M., Galloway, L., & El Ghak, T. (2020). Entrepreneurial orientation, access to financial resources and SMEs' business performance: The case of the United Arab Emirates. *The Journal of Asian Finance, Economics and Business*, *7*(12), 465-474.

PART-3

Regarding the long-term success of ChatGPT, the company's commitment to both business and technological growth is essential given that it is an AI language model. A study was conducted to determine the best course of action for ChatGPT and how it can be put to use to guarantee the company's growth and survival. The research looked into what factors contribute to ChatGPT's success and how companies may implement the tool to boost customer satisfaction, streamline operations, and increase income. The research also highlighted the need for constant development of ChatGPT to adapt to changing client needs and technical progress. Overall, the study shows how crucial ChatGPT is in defining the firm's future, and how its growth and durability may be ensured through constant innovation and integration into existing business procedures.

3.1. SWOT analysis of the study

The SWOT figure shows that ChatGPT can be a tool for entrepreneurs to explore the possibilities and difficulties of AI in specific industries while also enabling industry-specific business growth and sustainability. The title is contemporary and pertinent, given the continued importance of technology in entrepreneurship. Still, it should also consider AI's limitations and potential adverse effects in these sectors.

Strengths

- Contemporary and emerging issue
- Multidisciplinary approach
- Interdisciplinary focus
- Perceived Usefulness centered

Weaknesses

- No/Little concern
- Bias of means

Opportunity	Threats
• New academic buzzword	• Competition
• Look forward to new tools (Bard)	• Data limitations

The research paper, on its whole, possesses several strengths and opportunities, including the chance to contribute to academic knowledge and a concentration on a relevant and timely topic. One must also take into account the research's potential for bias as well as its limited scope.

3.2. Future Research: Emerging of Entrepreneurial research through ChatGPT

Theories can be used for ChatGPT and entrepreneurial research

Technology Acceptance Model (TAM) can be applied in both ChatGPT and entrepreneurial research. In ChatGPT research, TAM can help researchers understand the factors that influence user acceptance and use of the chatbot, including perceived usefulness and perceived ease of use. By addressing these factors, researchers can improve the chatbot's performance and drive user adoption. In entrepreneurial research, TAM can help entrepreneurs identify the factors that drive the adoption of new technologies, including the perceived usefulness and ease of use of the technology, compatibility with existing processes, and external pressures. By understanding these factors, entrepreneurs can make informed decisions about technology adoption to gain a competitive advantage.

TTF (Task-Technology Fit) theory provides a framework for understanding the relationship between technology and tasks within a domain. In the context of ChaTGPT, the theory can help to assess how well the language model fits with the various tasks and processes in the domains it is being used in, such as customer support, fraud detection, employee collaboration, bookkeeping, and patient-doctor communication. The TTF theory suggests that a good fit between technology and tasks results in higher levels of task performance and user satisfaction. By applying the TTF framework, a study could assess the extent to which ChaTGPT supports these tasks and identify areas for improvement.

HCI (Human-Computer Interaction) theory provides a framework for understanding the interactions between people and technology. This theory can help to assess the user experience of ChaTGPT in the domains mentioned above, including the ease of use, level of user satisfaction, and impact on human behavior and cognition. For example, the study could assess how well users are able to interact with ChaTGPT and how the language model affects their work processes, as well as their overall satisfaction with the technology.

The TOE (Technology-Organization-Environment) framework provides a comprehensive perspective on the impact of technology on organizations. The framework suggests that technology affects the organization as a whole, including the organizational structure, culture, and processes. In the context of ChaTGPT, a study using the TOE framework could assess the impact of the language model on the organization, such as how it affects the organizational structure and the culture of the organization. For example, the study could assess whether ChaTGPT has led to changes in the way work is organized or whether it has impacted the organization's values and beliefs.

Finally, the Diffusion of Innovation (DOI) theory explains how and why new ideas and technologies spread within a social system. This theory can be applied to the study of the use of ChaTGPT in business, management, entrepreneurship, and other operations to understand the factors that influence its adoption and diffusion. For example, the study could assess how ChaTGPT is being adopted and diffused within organizations, and what factors are driving or hindering its adoption and diffusion. The study could also identify opportunities to increase the diffusion and adoption of the language model.

In conclusion, by applying these theories, a study could provide a comprehensive understanding of the use of ChaTGPT in various domains, including the factors that influence its adoption, the impact it has on organizations and users, and opportunities for improvement.

3.3. Methodologies can be used for ChatGPT and entrepreneurial research

The methodology for this study could involve a combination of qualitative and quantitative research methods. The following steps could be taken:

- ***Literature review:*** This step would involve reviewing existing literature on ChatGPT, SMEs, entrepreneurial firms, and sustainability, with a focus on their relationship to venture creation in the AI revolution. This would help to identify the current state of knowledge on the topic and identify any gaps that the study can fill.
- ***Case studies:*** Case studies could be conducted of SMEs and entrepreneurial firms that have successfully utilized ChatGPT and other AI technologies for venture creation. These case studies could provide valuable insights into the challenges and opportunities that these companies have faced and the impact of AI on their business.
- ***Surveys and questionnaires:*** Surveys and questionnaires could be used to gather data from a larger sample of SMEs and entrepreneurial firms. The surveys could explore the use of AI technologies, including ChatGPT, for venture creation and its impact on business sustainability.
- ***Expert interviews:*** Expert interviews could be conducted with leading figures in the fields of AI, entrepreneurship, and sustainability. These interviews could

provide additional insights into the implications of ChatGPT and other AI technologies for venture creation and sustainability.

The use of a mixed-methods approach, combining both qualitative and quantitative methods, would provide a more comprehensive understanding of the relationship between ChatGPT, SMEs, entrepreneurial firms, and sustainability and their implications for venture creation in the AI revolution. The data collected from each method could be triangulated to validate the findings and strengthen the conclusions.

3.4. Other areas of future study

Though, ChatGPT is playing a significant role in various various business and economic domain, including but not limited to management, finance, accounting, marketing, entrepreneurship, supply chain management, industry, human resource management, innovation, customer service, corporate governance, leadership, project management, green business practices, education, medical, programming, healthcare, research and so on. Other areas of future study based on the study

- Add extension of voice or speech recognition.
- Advancing AI research including feelings, emotions.
- Use ChatGPT to develop a robot, who can help human being in preparing breakfast.
- Think, at the morning your robot is waiting for my breakfast.
- Advancing the tool by introducing demographic information
- Enhance the quality of support.
- Continuous process until it comes 0.
- Further study needed Add more recent years 2022, 2023 data
- Potential error in command may hamper the significance of the meaning

3.5. Limitations

The study had some limitations, such as the limited availability of data and the subjective nature of the data collected. Additionally, the study relied solely on document analysis, which may have resulted in a limited sample size and may not have captured the full experiences and perspectives of entrepreneurs and SME owners. However, the findings of the study provide valuable insights into the challenges and benefits of using AI and ChatGPT for entrepreneurs and SME owners and can be used as a basis for future research.

A. Main Limitation is data store, it has preserved the data till 2021. But we are living 2023. How accurate the answer can be?
B. Answers based on the previous chat, cannot think out of thebox
C. All of the answers based on bullet points number
D. Server slow.

PART-4

Google's Bard vs. OpenAI's ChatGPT: AI Championship and Social Change through 4IR

Md. Saidur Rahaman, M. M. Tahmid Ahsan, Md. Mizanur Rahman, & Nishath Anjum

4.1. Introduction

Through the Fourth Industrial Revolution, the introduction of powerful AI technologies like Google's Bard and OpenAI's ChatGPT has the potential to bring about major social change (4IR). The fast progress being made in the field of artificial intelligence and the heated competition between industry leaders are both on display in the AI championship between these two systems. The use of these AI tools may lead to greater output and efficiency across several sectors. These systems let people to devote their time and energy to more complicated and imaginative endeavours by automating routine operations. This has the potential to spur more creative solutions and better consumer goods. But, AI could also have certain unintended consequences for the human race. When more work is automated, one worry is that people will lose their jobs. Because of this, it is possible that some groups of employees may be hit worse than others, further widening the gap between them. Furthermore, as AI systems collect and analyse enormous volumes of data, there are privacy and security concerns. The ethical and social consequences of new technologies must be taken into account if their benefits are to be fully realised and their potential dangers mitigated. One aspect of this is making sure that all people are considered fairly and equitably in the creation and implementation of AI systems. Furthermore, it includes addressing any biases in the data and algorithms utilised by these systems, as well as ensuring that users are held accountable for their interactions with AI. And the potential effects of AI on the future of labour and the economy must be taken into account. This includes looking for new options for workers who may be displaced from their jobs and for providing aid to such workers. In conclusion, the AI competition between Google's Bard and OpenAI's ChatGPT demonstrates the potential for AI to bring about major social change through the 4IR. Yet, it is essential to think about the ethical and social implications of new technologies and endeavour to guarantee that they are built upon a foundation of fairness, transparency, and accountability.

4.2. Background of the research

Artificial Intelligence refers to an area of computer science that focuses on creating machines that can perform tasks that typically require human intelligence (Dellermann et al., 2019). The common usages of AI include understanding natural language, recognizing any object in pictures and making decisions effectively for the ease of the users. Among the branches of AI, Natural Language Processing (NLP) is a significant one. (Lauriola et al., 2022) It normally deals with the interactions between computerized systems and humans using natural language. The AI language models are generally NLP based technology where the NLP works as the fuel for this to run the activities (Luitse & Denkena,2021). Generally it uses statistical learning techniques to process and generate text replies like the human beings. These models are specially trained on generating text responses, making things easier for the users, handling a large number of data with efficiency, processing a huge amount of data with efficiency and tasks like this.

Artificial intelligence is a branch of the computer science and engineering sector, focusing on the things that can be performed without the help of human beings. (Jordan, 2019). It can perform the activities with efficiency of human level but with much quicker time.

The language models has made significant progress in recent years. The reasons behind this progress are like the advancement in deep learning techniques, increase in the computational ability and also, the emphasis given by the tech giants into this sector. At present, two of the most powerful language models are the ones namely Google's BARD and OpenAI'sChatGPT. These models have all the capabilities of becoming a revolution in this sector and change the landscape of this area of NLP. This NLP models also have some practical implications in the areas like marketing, education, finance, healthcare, customer services, banking, software and many other sectors.

It is very clear that both the Bard (Google) and the ChatGPT (OpenAI) are the most advanced NLP models (Hegde, 2023). They have a very significant implications for a wide variety of industries and applications. Before starting the comparison part, it should be mentioned that these two models have started off a championship race between themselves for winning the throne and both of them are prepared to show the best of the NLP displays in order to win it. These two models have the capabilities of becoming the best in this competition, still there is a long way to go for both of them. However, beyond the technological competition, the success of these models also highlights the need for responsible AI development and deployment for the welfare of the society as well as for the mankind. And also, they have some more duties to make sure that the ethical standards are maintained and no unethical activities are getting promoted or benefitted due to this implication of NLP.

The aim of producing this paper is to make a comparison and evaluation of the performance of two state-of-the-art language models: Google's BARD and OpenAI'sChatGPT. The basis for comparison will be various performance metrics such as accuracy, fluency, and consistency. Moreover, their contributions to the society will also be observed. The paper aims to find out the answers to the following research questions:

1. How can BARD and ChatGPT be compared in terms of performance on various NLP tasks?

2. What are the strengths of these models? Also, what are the weaknesses of each of these models, and how can these models complement each other?

3. What are the core implications towards social change through 4IR of these findings for the future of AI and NLP?

The ultimate objective of this paper us to provide with a comprehensive and objective comparison and evaluation of these two language models and to find out who would be on the upper hand in terms of getting a stronger position in the marketplace. Moreover, evaluating the contribution of these models to the advancement of the field of NLP. The findings of this study will be a very helpful source of information for the practitioners and researchers who are interested in implementing AI systems for NLP applications. They would be able to follow the procedures of preparing this paper and also, the methodology of preparing a paper regarding this particular sector. They would also have the relevant guidance regarding the impacts and influences that AI championship battle will have on the technology world as well as in the world economy.

4.3. Google's Bard and OpenAI's ChatGPT: Overview

BARD refers to a large, pre-trained NLP based language model developed by Google Inc. (Duran et al., 2021). The basic of this model is transformer architecture, which is a neural network, widely renowned for performing the NLP tasks. BARD is pre-trained on handling a huge amount of text data (Han et. Al., 2021). This allows it to perform well on a wide range of NLP tasks without the need for much of the fine-tuning. Google has trained BARD on a diverse set of texts and natural languages. Therefore, it can generate coherent and fluent text in accordance with the input. It provides the user with the type of result he or she wants in a lightning speed.

On the contrary, OpenAI's invention ChatGPT is also a pre-trained language model. This model is basically based on the transformer architecture. It was trained on a large corpus of text data from the internet, including websites, forums, and social media platforms. ChatGPT is designed for conversational AI and can generate human-like text that is both accurate and fluent. It has been used for a variety of tasks such as question answering, text completion, and text generation.

Both BARD and ChatGPT are among the largest language models available today and have shown remarkable performance on various NLP tasks. However, their training objectives and target applications are different, and it would be interesting to compare their performance and see how they complement each other.

4.4. State-of-the-art in AI and language models

Artificial Intelligence, which is commonly renowned as AI and language models have experienced a really rapid growth in recent years. The advancement of deep learning techniques is one of the major reasons behind that. Also, due to the increasing availability of large amounts of text data, language models have become one of the most active and attractive areas of research and development in NLP.

The current state-of-the-art in language models are the Bard by Google, which has been ran by LaMDA, and OpenAI'sChatGPT, which has been based on the GPT (Generative Pretrained Transformer) architecture. Both these models are pre-trained on handling and processing immersive amounts of data in the form of text and also, carry out the best possible results for the users. These models can also be fine-tuned for some specific NLP tasks. These tasks can be exemplified by text classification, answering to the questions, generating text according to the command.

The success that these pre-trained language models have gained over a short period of time has led the focus of the manufacturers in building up even larger models. The growing demand for the bigger models has been a joyful thing for the programmers who work in this sector. For example, OpenAI'sChatGPT run by the architecture GPT-3. It has over 175 billion parameters. This has made the ChatGPT one of the largest language models available today. These models have shown remarkable performance on a variety of NLP tasks and have the potential to revolutionize the field of NLP.

However, with the increased size and complexity of these models, there are also concerns about their ethical and social implications, such as the potential for bias in their predictions and the energy consumption required for training and running these models. The state-of-the-art in AI and language models is therefore a rapidly evolving field, with ongoing research and development aimed at both improving performance and addressing these concerns.

4.5. Previous research on BARD and ChatGPT

There has been limited number of research comparing the performance of Google's BARD and OpenAI'sChatGPT. The reason behind that Google's Bard is a relatively new model. However, both models have been extensively studied individually. But still there will be many research conducted in the coming months.

Research on Bard has the main focus on the evaluation of the performance of Bard on various NLP tasks. These tasks include text classification, sentiment analysis, and question answering. Google has presented that Bard can perform BARD has been shown to perform well on these tasks and to outperform other AI based NLP based models. The research has also explored the feature of interpretability of BARD, demonstrating and explaining the ability it has to capture the underlying meaning of text. The spotlight of the research also was also on the ability of Bard on generating coherent and meaningful text.

Research on ChatGPT has focused on evaluating its performance as a conversational AI system. Studies have shown that ChatGPT is able to generate human-like text that is both accurate and fluent, making it a promising tool for a variety of conversational AI applications, such as customer service and virtual assistants. The research has also explored to find out the limitations that ChatGPT has, such as the potential for bias in its predictions and the limitations of its training data.

Given the limited research comparing the performance of BARD and ChatGPT, this paper aims to fill this gap by providing a comprehensive and objective evaluation of these two models. The results of the study will contribute to a deeper understanding of the strengths and weaknesses of these models and their implications for the future of AI and NLP.

Previous research on BARD and ChatGPT has produced the following key findings:

1. It has been shown by Google that Bard is capable of performing a wide range of NLP tasks at a really high level of efficiency. Some of these tasks are really complex and difficult to perform in such a quick time. These tasks include the tasks like text classification, sentiment analysis, and question answering.
2. Bard has a bidirectional nature and also, it has the ability to capture context. This has been demonstrated to be key factors in its performance.
3. Bard has been found to be really effective in capturing the underlying meaning of text, making it useful for tasks such as text classification and sentiment analysis.
4. The interpretability of Bard has been studied, and it has been shown by Google that Bard is able to generate coherent and meaningful text.

4.6. ChatGPT

1. ChatGPT has been shown to generate human-like text that is both accurate and fluent. Which makes it a really promising tool in the arena of conversational AI applications.
2. ChatGPT has been found to have some shortcomings as they are limited by the biases and limitations in training the data, sometimes unintentionally it leads to the potential for biased predictions.
3. The limitations of ChatGPT's training data have also been shown to limit its ability to generalize to new domains and situations.
4. Research has also put a spotlight on the area of energy consumption required for training and running ChatGPT.

These findings would provide with a strong and concrete foundation for the current study. This would really be helpful to compare and evaluate the performance of Bard and ChatGPT. This comparison will also help in making a comprehensive and objective evaluation of these two rival models. The results of the study will also have a significant contribution to a deeper understanding of the strengths and weaknesses of these models. Also the analysis will help in measuring and discussing their implications for the future of AI and NLP.

4.7. Methodology

Research design

The research design for the "Google's BARD vs. OpenAI'sChatGPT: AI Championship" paper consists of the following steps:

1. Data Collection: This step includes collection of a large set of diverse and representative dataset of texts in various domains and languages. It is one of the most significant steps in the whole research.

2. Pre-processing: Clean and pre-process the collected text data with a view to remove any irrelevant information (if there is any). Also ensuring consistency across the data collected.

3. Model Selection: Google's Bard and OpenAI'sChatGPT have been selected as the models to be compared and evaluate their performance on the selected NLP tasks.

4. Task Preparation: In this step, the NLP tasks are prepared for evaluation, such as text classification, analyzing the sentiments which is also known as the sentiments analysis, and question answering.

5. Model Training and Evaluation: Train and fine-tune both the models (Bard and ChatGPT) models on the pre-processed data, and also evaluate their performance on the NLP tasks.

6. Result Analysis: Analyze the results of the model evaluation, including accuracy, speed, and interpretability, and compare the performance of Bard and ChatGPT on the standards of NLP tasks.

7. Discussion: This part contains the discussions about the implications of the tools, including any strengths and weaknesses of the models, their potential for real-world applications, and the future directions for research in AI and NLP.

This research design will allow for a comprehensive and objective evaluation of BARD and ChatGPT, and provide a basis for comparison and analysis of their performance on a range of NLP tasks. The results of the study will contribute to a deeper understanding of the strengths and weaknesses of these models and their implications for the future of AI and NLP.

4.8. Data collection

Data collection has always been one of the core activities of the research and the collected data makes the researcher able to enter to the next step.

One of the most crucial and critical steps in the research design for the "Google's Bard vs. OpenAI'sChatGPT: AI Championship" paper is collecting the necessary data for conducting the research. The following steps can be helpful in the process of collecting the data:

1. Domain Selection: Selection of a range of text domains, such as news articles, social media posts, and customer reviews, to ensure that the data is representative and diverse as well as relevant to the study.

2. Data Sources: Identify and collect data from appropriate and relevant sources, these sources include online news websites, social media platforms, and customer review websites.

3. Data Filtering: Filter the data that has been collected in order to remove any irrelevant or information of below per value and to ensure that the data has the consistency and also, enough relevance with the study.

4. Data Annotation: Annotate the data for the purpose of indicating the categories or sentiments represents, if relevant, such as any positive or negative sentiment or news articles about politics or technology.

5. Data Splitting: Splitting the data that has been collected and annotated data into training, validation, and testing sets, ensuring that the data is representative of the entire dataset.

The collected data has to be diverse and representative, covering a range of text domains and languages, and should be pre-processed in order to ensure the consistency and quality. The data should also have be split into appropriate sets for training, validation, and testing. These activities would ensure that the results of the model evaluation are robust and accurate.

4.9. Data analysis

Data analysis is one of the most impactful activities in the whole research as the most crucial thing of the research, the findings depends on the data analysis part and if the data analysis is not done perfectly, then it would definitely affect the overall findings of the research. And therefore, a wrong message will be delivered through the research.

Data analysis is a very important step in the research design for the "Google's BARD vs. OpenAI'sChatGPT: AI Championship" paper, where the results of the model evaluation are analyzed and compared. The following steps can be taken to analyze the data:

1. Model Evaluation Metrics: Choose appropriate evaluation metrics, such as accuracy, F1 score, and recall, to measure the performance of the models on the NLP tasks.

2. Result Comparison: Compare the results of the model evaluation for BARD and ChatGPT on the selected NLP tasks, and create visualizations, such as bar charts or line graphs, to present the results.

3. Error Analysis: Analyze the errors made by the models, including any patterns or trends in the types of errors made, and identify any factors that may have contributed to the errors.

4. Model Interpretability: Evaluate the interpretability of the models, such as the ability to understand the reasoning behind their predictions, and compare the interpretability of BARD and ChatGPT.

5. Model Comparison: Compare the overall performance of BARD and ChatGPT, including accuracy, speed, and interpretability, and identify any strengths and weaknesses of each model.

The data analysis should provide with a comprehensive and objective comparison and evaluation of the performance of BARD and ChatGPT on the NLP tasks and should provide a basis for comparing and contrasting the models. The results of the data analysis should also highlight any areas for improvement or further research in AI and NLP.

4.10. Model Evaluation Metrics

Accuracy

OpenAI's GPT-3 (ChatGPT) and Google's Bard are both state-of-the-art models in their respective fields, but they are developed by the developers with a view to serve different purposes and are optimized for different tasks. Bard is a pre-trained language representation model fine-tuned for specific NLP tasks such as sentiment analysis and question answering, and has shown high accuracy in those tasks. On the other hand, GPT-3 is a large language generation model, trained to generate coherent and contextually appropriate text. The accuracy of these models depends on the specific tasks and data that they are being applied to, and a direct comparison of their accuracy is not straightforward.

F1 Score

In summary, it's really difficult thing the F1 scores of Google Bard and OpenAI'sChatGPT, it's difficult to compare the F1 scores of Google's Bard and OpenAI's GPT-3 as they are optimized for performing different types of tasks. Google's Bard is a pre-trained language representation model. It is fine-tuned especially for performing some specific Natural Language Processing (NLP) tasks such as sentiment analysis and question answering, and has shown high F1 scores in those tasks. On the other hand, GPT-3 is a large language generation model that is not designed to be fine-tuned for specific NLP tasks, and its quality is typically evaluated based on human judgment and its ability to generate coherent and contextually appropriate text, rather than on F1 score. The F1 score of a model will depend on the specific task and data it is being applied to, so the better model will depend on the requirements of the specific task at hand.

Recall

In short, it can be stated that and GPT-3 have different architectures and are optimized for different tasks, so it's not appropriate to compare their recall capabilities directly. BERT is a pre-trained language representation model fine-tuned for specific NLP tasks, while GPT-3 is a large language generation model. Recall in the context of BERT refers to its ability to recall specific information from its training data, while GPT-3 is not designed to recall specific information but to generate text based on a given prompt. The recall capability of a model depends on the specific task and data it is being applied to, and the better model will depend on the requirements of the specific task at hand.

4.11. Comparison of NLP tasks between Bard and ChatGPT

Both the Google's Bard and OpenAI's ChatGPT-3 are both advanced models in the arena of Natural Language Processing (NLP), but they are optimized for performing different types of Natural Language Processing (NLP) tasks.

Bard is a pre-trained language representation model that has been fine-tuned for performing specific NLP tasks such as sentiment analysis, entity recognition, and question answering. Bard's pre-training and fine-tuning process allows it to effectively capture the relationships between words in a sentence, which is important for these NLP tasks. And this thing makes Bard a very strong player in this sector.

On the other hand, GPT-3 is a large language generation model that has been trained to generate coherent and contextually appropriate text based on a given prompt. It has the ability to perform a wide range of NLP tasks such as text summarization, translation, and conversation, but its main strength is in its ability to generate text.

In general, the performance of a model on specific NLP tasks will depend on some specific factors like the quality and quantity of the training data, the architecture of the model, and the specific fine-tuning process. The better model for a specific NLP task will depend on the requirements of the task at hand.

4.12. Discussion

The AI championship between Google's Bard and OpenAI's ChatGPT illustrates the potential for powerful AI technology to bring about major social change through the implementation of the 4IR. Yet, in order to ensure that the benefits of AI are maximised while the hazards are reduced, it is vital to evaluate the ethical and societal implications of these technologies. These consequences include the possibility of job displacement as well as concerns over privacy. In order to accomplish this, the creation and deployment of AI systems must be conducted with complete openness and responsibility, as well as a firm dedication to the principles of fairness and equality.

The role of technology in social change refers to the impact that technology has on society and its cultural, political, and economic structures. Technology has been a driving force in shaping human history and has had a profound impact on the way people live, work, and interact with each other. The widespread adoption of technology has led to significant social changes, including increased globalization, the rise of digital communication and information, and the transformation of traditional industries. The impact of technology on social change is complex and multi-faceted, and it is essential to understand its effects in order to guide its future development and use.

The purpose of the study on OpenAI's ChatGPT is to explore the potential impact that AI language models like ChatGPT may have on social change. ChatGPT is a cutting-edge AI technology that has the ability to generate human-like text, which has numerous applications in areas such as customer service, content creation, and education.

The significance of this study lies in its potential to contribute to the understanding of how AI technology like ChatGPT can influence social change. By conducting a qualitative study on OpenAI's ChatGPT, the researchers aim to gain insights into the ways in which AI language models may shape our future society and the potential implications of their widespread adoption. The findings of this study will be of interest to a wide range of stakeholders, including researchers, policymakers, industry experts, and the general public, who are interested in the impact of technology on society.

Conclusion

In conclusion, the Fourth Industrial Revolution has dramatically advanced by creating cutting-edge artificial intelligence (AI) technologies like Google's Bard and OpenAI's ChatGPT (4IR). These AI systems can potentially alter our civilizations as they develop and improve significantly. The duel between OpenAI's ChatGPT and Google's Bard in the AI championship exemplifies the industry leaders' rivalry and the quick development of AI technology. The development made in this area is encouraging since it can result in increased output, better judgement, and more effective use of resources. As with every technological development, there may be societal repercussions to take into account. While AI has the potential to impact society positively, it also poses a threat to privacy and security, can cause job displacement, and exacerbates inequality. It is crucial to ensure that ethical considerations and a commitment to social responsibility inform AI development and implementation as we explore its potential. Doing this allows us to use AI's disruptive power to improve everyone's future.

References

Dellermann, D., Ebel, P., Söllner, M., & Leimeister, J. M. (2019). Hybrid intelligence. *Business & Information Systems Engineering*, *61*, 637-643.

Duran, N., Battle, S., & Smith, J. (2021). Sentence encoding for Dialogue Act classification. Natural Language Engineering, 1-30. doi:10.1017/S1351324921000310

Han, X., Zhang, Z., Ding, N., Gu, Y., Liu, X., Huo, Y., ... & Zhu, J. (2021). Pre-trained models: Past, present and future. *AI Open*, *2*, 225-250.

Hegde, V.V. (2023), Why Google's Bard AI Lost to ChatGPT, DEV, Access from, https://dev.to/varshithvhegde/why-googles-bard-ai-lost-to-chatgpt-5f2k

Jordan, M. I. (2019). Artificial Intelligence—The Revolution Hasn't Happened Yet. Harvard Data Science Review, 1(1). https://doi.org/10.1162/99608f92.f06c6e61

Lauriola, I., Lavelli, A., & Aiolli, F. (2022). An introduction to deep learning in natural language processing: Models, techniques, and tools. *Neurocomputing*, *470*, 443-456.

Luitse, D., & Denkena, W.. (2021). The great Transformer: Examining the role of large language models in the political economy of AI. *Big Data & Society*, *8*(2), 205395172110477. https://doi.org/10.1177/20539517211047734

Book Cover page designer: Nasim Nawaz Md Shuvo, 171-116-099, BBA 48th, Major in HRM, Department of Business Administration, Metropolitan University Sylhet, Bangladesh.

Publisher: Eliva Press Global Ltd

Email: info@elivapress.com

Made in the USA
Middletown, DE
05 November 2026

36562490R00056